## Nature's Formations

# How Are Lakes Formed?

B.J. Best

New York

Published in 2018 by Cavendish Square Publishing, LLC
243 5th Avenue, Suite 136, New York, NY 10016

Copyright © 2018 by Cavendish Square Publishing, LLC

First Edition

No part of this publication may be reproduced, stored in a retrieval system, or transmitted in any form or by any means—electronic, mechanical, photocopying, recording, or otherwise—without the prior permission of the copyright owner. Request for permission should be addressed to Permissions, Cavendish Square Publishing, 243 5th Avenue, Suite 136, New York, NY 10016. Tel (877) 980-4450; fax (877) 980-4454.

Website: cavendishsq.com

This publication represents the opinions and views of the author based on his or her personal experience, knowledge, and research. The information in this book serves as a general guide only. The author and publisher have used their best efforts in preparing this book and disclaim liability rising directly or indirectly from the use and application of this book.

CPSIA Compliance Information: Batch #CS17CSQ

All websites were available and accurate when this book was sent to press.

Library of Congress Cataloging-in-Publication Data

Names: Best, B.J.
Title: How are lakes formed? / B.J. Best.
Description: New York : Cavendish Square Publishing, 2018. | Series: Nature's formations | Includes index.
Identifiers: ISBN 9781502625496 (pbk.) | ISBN 9781502625519 (library bound) | ISBN 9781502625502 (6 pack) | ISBN 9781502625526 (ebook)
Subjects: LCSH: Lakes--Juvenile literature.
Classification: LCC GB1603.8 B47 2018 | DDC 551.48'2--dc23

Editorial Director: David McNamara
Copy Editor: Nathan Heidelberger
Associate Art Director: Amy Greenan
Designer: Alan Sliwinski
Production Coordinator: Karol Szymczuk
Photo Research: J8 Media

The photographs in this book are used by permission and through the courtesy of: Cover Pung/Shutterstock.com; p. 5 Jeffrey M. Frank/Shutterstock.com; p. 7 Bob Coffen/Shutterstock.com; p. 9 Thom Lang/Corbis Documentary/Getty Images; p. 11 Andreea Dragomir/Shutterstock.com; p. 13 Andreas Strauss/LOOK-foto/Getty Images; p. 15 Ammit Jack/Shutterstock.com; p. 17 Pierre Leclerc/Shutterstock.com; p. 19 Brisbane/Shutterstock.com; p. 21 Gerald Bernard/Shutterstock.com.

Printed in the United States of America

# Contents

How Lakes Are Formed.......**4**

New Words...................... **22**

Index .............................. **23**

About the Author............. **24**

Lakes are in many places on Earth.

Some are small.

Some are huge!

Lakes are like bowls in the ground.

The bowls are called **basins**.

The basins are filled with water.

Some lakes were made by **glaciers**.

Glaciers are very thick sheets of ice.

9

Glaciers are big and heavy.

They covered the land.

They changed the land under them.

The glaciers melted.

Some basins were left.

The melting ice filled the basins.

This made lakes!

Some mountains have melted rock inside them.

These are called **volcanoes**.